Climate Change and National Security

An Agenda for Action

Joshua W. Busby

CSR NO. 32, NOVEMBER 2007
COUNCIL ON FOREIGN RELATIONS

The Council on Foreign Relations is an independent, nonpartisan membership organization, think tank, and publisher dedicated to being a resource for its members, government officials, business executives, journalists, educators and students, civic and religious leaders, and other interested citizens in order to help them better understand the world and the foreign policy choices facing the United States and other countries. Founded in 1921, the Council carries out its mission by maintaining a diverse membership, with special programs to promote interest and develop expertise in the next generation of foreign policy leaders; convening meetings at its headquarters in New York and in Washington, DC, and other cities where senior government officials, members of Congress, global leaders, and prominent thinkers come together with Council members to discuss and debate major international issues; supporting a Studies Program that fosters independent research, enabling Council scholars to produce articles, reports, and books and hold roundtables that analyze foreign policy issues and make concrete policy recommendations; publishing *Foreign Affairs*, the preeminent journal on international affairs and U.S. foreign policy; sponsoring Independent Task Forces that produce reports with both findings and policy prescriptions on the most important foreign policy topics; and providing up-to-date information and analysis about world events and American foreign policy on its website, CFR.org.

THE COUNCIL TAKES NO INSTITUTIONAL POSITION ON POLICY ISSUES AND HAS NO AFFILIATION WITH THE U.S. GOVERNMENT. ALL STATEMENTS OF FACT AND EXPRESSIONS OF OPINION CONTAINED IN ITS PUBLICATIONS ARE THE SOLE RESPONSIBILITY OF THE AUTHOR OR AUTHORS.

Council Special Reports (CSRs) are concise policy briefs, produced to provide a rapid response to a developing crisis or contribute to the public's understanding of current policy dilemmas. CSRs are written by individual authors—who may be Council Fellows or acknowledged experts from outside the institution—in consultation with an advisory committee, and are intended to take sixty days from inception to publication. The committee serves as a sounding board and provides feedback on a draft report. It usually meets twice—once before a draft is written and once again when there is a draft for review; however, advisory committee members, unlike Task Force members, are not asked to sign off on the report or to otherwise endorse it. Once published, CSRs are posted on the Council's website, CFR.org.

For further information about the Council or this Special Report, please write to the Council on Foreign Relations, 58 East 68th Street, New York, NY 10065, or call the Communications office at 212-434-9888. Visit our website, CFR.org.

To submit a letter in response to a Council Special Report for publication on our website, CFR.org, you may send an email to CSReditor@cfr.org. Alternatively, letters may be mailed to us at: Publications Department, Council on Foreign Relations, 58 East 68th Street, New York, NY 10065. Letters should include the writer's name, postal address, and daytime phone number. Letters may be edited for length and clarity, and may be published online. Please do not send attachments. All letters become the property of the Council on Foreign Relations and will not be returned. We regret that, owing to the volume of correspondence, we cannot respond to every letter.

CONTENTS

FOREWORD

Climate change presents a serious threat to the security and prosperity of the United States and other countries. Recent actions and statements by members of Congress, members of the UN Security Council, and retired U.S. military officers have drawn attention to the consequences of climate change, including the destabilizing effects of storms, droughts, and floods. Domestically, the effects of climate change could overwhelm disaster-response capabilities. Internationally, climate change may cause humanitarian disasters, contribute to political violence, and undermine weak governments.

In this Council Special Report, Joshua W. Busby moves beyond diagnosis of the threat to recommendations for action. Recognizing that some climate change is inevitable, he proposes a portfolio of feasible and affordable policy options to reduce the vulnerability of the United States and other countries to the predictable effects of climate change. He also draws attention to the strategic dimensions of reducing greenhouse gas emissions, arguing that sharp reductions in the long run are essential to avoid unmanageable security problems. He goes on to argue that participation in reducing emissions can help integrate China and India into the global rules–based order, as well as help stabilize important countries such as Indonesia. And he suggests bureaucratic reforms that would increase the likelihood that the U.S. government will formulate effective domestic and foreign policies in this increasingly important realm.

The result is an authoritative, well-written, and practical paper that merits careful consideration by members of Congress, the administration, and other interested parties in the United States and internationally.

Richard N. Haass
President
Council on Foreign Relations
November 2007

ACKNOWLEDGMENTS

In developing this Council Special Report, I interviewed a number of individuals who work on climate change, national security, and the intersection between the two. These included current and former U.S. government officials, former members of the military, and academics, as well as staff from international organizations, nongovernmental organizations, and businesses. During the course of writing this report, I consulted with an advisory group that met to offer constructive feedback. I am grateful to Kurt M. Campbell for chairing the advisory committee and to Kent Hughes Butts, Helima L. Croft, John Gannon, Lukas Haynes, Paul F. Herman Jr., Jeff Kojac, Marc A. Levy, Meg McDonald, Alisa Newman Hood, Stewart M. Patrick, Joseph Wilson Prueher, Nigel Purvis, P.J. Simmons, and R. James Woolsey for participating. I would also like to thank Shannon Beebe for his helpful comments and advice on this project.

I thank Council President Richard N. Haass for his support in producing this CSR. I thank Vice President and Director of Studies Gary Samore for his helpful suggestions. I am grateful for the advice and support of Sebastian Mallaby, director of the Maurice R. Greenberg Center for Geoeconomic Studies, and Michael A. Levi, director of the Program on Energy Security and Climate Change. I also thank the Publications team of Patricia Dorff and Lia Norton and the Communications team headed by Lisa Shields and Anya Schmemann. This publication was sponsored by the Geoeconomics Center and was made possible, in part, by a grant from the John D. and Catherine T. MacArthur Foundation. The statements made and views expressed in this report are solely my responsibility.

Joshua W. Busby

COUNCIL SPECIAL REPORT

INTRODUCTION

When Hurricane Katrina struck New Orleans in 2005, Americans witnessed on their own soil what looked like an overseas humanitarian-relief operation. The storm destroyed much of the city, causing more than $80 billion in damages, killing more than 1,800 people, and displacing in excess of 270,000. More than 70,000 soldiers were mobilized, including 22,000 active duty troops and 50,000-plus members of the National Guard (about 10 percent of the total Guard strength). Katrina also had severe effects on critical infrastructure, taking crude oil production and refinery capacity off-line for an unprecedented length of time. At a time when the United States was conducting military operations in Afghanistan and Iraq, the country suddenly had to divert its attention and military resources to respond to a domestic emergency.

Climate change and Katrina cannot be linked directly with each other, but the storm gave Americans a visual image of what climate change—which scientists predict will exacerbate the severity and number of extreme weather events—might mean for the future.[1] It also began to alter the terms of the climate debate. The economics community has been engaged in an important, ongoing discussion since the early 1990s about whether early action to prevent climate change is justified; this debate has compared the potential economy-wide costs of lowering greenhouse gas emissions to the possible economic costs of climate change. In 2007, the debate turned, broadening beyond economics to include, in particular, the consequences of climate change for national security. In March 2007, Senators Richard J. Durbin (D-IL) and Chuck Hagel (R-NE) introduced a bill requesting a National Intelligence Estimate to assess whether and how climate change might pose a national security threat. In April 2007, the CNA Corporation, a think tank funded by the U.S. Navy, released a report on climate change

---

[1] Scientists do not attribute single weather events like Katrina to climate change; at most, they would say that climate change make extreme storms like Katrina more likely. Whether climate change has been responsible for an increase in both the severity and number of hurricanes is one of the most hotly debated subjects in the scientific community.

and national security by a panel of retired U.S. generals and admirals that concluded: "Climate change can act as a threat multiplier for instability in some of the most volatile regions of the world, and it presents significant national security challenges for the United States." That same month, the UN Security Council—at the initiative of the UK government—held its first-ever debate on the potential impact of climate change on peace and security. In October 2007, the Nobel committee recognized this emerging threat to peace and security by awarding former vice president Al Gore and the Intergovernmental Panel on Climate Change its peace prize. In November 2007, two think tanks, the Center for Strategic and International Studies (CSIS) and the Center for a New American Security (CNAS), released another report on the issue, concluding from a range of possible scenarios of climate change that, "We already know enough to appreciate that the cascading consequences of unchecked climate change are to include a range of security problems that will have dire global consequences."[2]

The new interest in climate change and national security has been a valuable warning about the potential security consequences of global warming, but the proposed solutions that accompanied recent efforts have emphasized broader climate policy rather than specific responses to security threats. Because the links between climate change and national security are worthy of concern in their own right, and because some significant climate change is inevitable, strategies that go beyond long-run efforts to rein in greenhouse gas emissions are required. This report sharpens the connections between climate change and national security and recommends specific policies to address the security consequences of climate change for the United States.

In all areas of climate change policy, adaptation and mitigation (reducing greenhouse gas emissions) should be viewed as complements rather than competing alternatives—and the national security dimension is no exception. Some policies will be targeted at adaptation, most notably risk-reduction and preparedness policies at home and abroad. These could spare the United States the need to mobilize its military later to rescue people and to prevent regional disorder—and would ensure a more effective response if such mobilization was nonetheless necessary. Others will focus on mitigation,

---

[2] CSIS/CNAS, *The Age of Consequences: The Foreign Policy and National Security Implications of Global Climate Change*, November 2007; available at http://www.cnas.org/climatechange.

which is almost universally accepted as an essential part of the response to climate change. Mitigation efforts will need to be international and involve deep changes in the world's major economies, such as those of China and India. As a result, the processes of working together to craft and implement them provide opportunities to advance American security interests. Such opportunities exist within many areas of climate policy: military-to-military workshops on emergency management, for example, can help other states deal with new security threats and, at the same time, cement strong relationships that can pay off in other national security dimensions.

EFFECTS OF CLIMATE CHANGE AND
CONSEQUENCES FOR U.S. NATIONAL SECURITY

The 2007 report from the Intergovernmental Panel on Climate Change (IPCC), the leading expert body in this field, summarizes the effects of climate change by kind, likelihood, and impact on different sectors such as agriculture and human health (see Table 1). Its main conclusion is that "some weather events and extremes will become more frequent, more widespread, and/or more intense during the 21st century."[3]

Table 1: Summary of Expected Effects in IPCC 2007 Report

Phenomenon and Direction of Trend	21st Century Likelihood
Over most land areas, warmer and fewer cold days and nights, warmer and more frequent hot days and nights	Virtually certain
Warm spells/heat waves. Frequency increases over most land areas	Very likely
Heavy precipitation events. Frequency increases over most areas	Very likely
Area affected by drought increases	Likely
Intense tropical cyclone activity increases	Likely
Increased incidence of extreme high sea level (excluding tsunamis)	Likely

Sources: IPCC Interim Working Group Report 1, April 2007; IPCC Synthesis Report, November 2007.

While some areas in northern Europe, Russia, and the Arctic may experience more positive effects of a warming climate in the short run, the long-run net consequences for all regions are likely to be negative if nothing at all is done to reduce emissions of greenhouse gases. Africa and parts of Asia are particularly vulnerable, given their locations and their limited governmental capacities to respond to flooding, droughts, and declining food production. Even the United States will face negative impacts from

---

[3] This report focuses on physical effects that scientists already regard as those most likely to surface in the coming decades, rather than more long-term, uncertain, or unlikely effects, which would include abrupt climate change and the scenario of a twenty-foot sea-level rise popularized in former U.S. vice president Al Gore's film *An Inconvenient Truth*.

droughts, heat waves, and storms. Each of these has potential consequences, direct and indirect, for national security.

National security extends well beyond protecting the homeland against armed attack by other states, and indeed, beyond threats from people who purposefully seek to damage or destroy states. Phenomena like pandemic disease, natural disasters, and climate change, despite lacking human intentionality, can threaten national security. For example, the 2006 U.S. National Security Strategy (NSS) notes that the Department of Defense has been charged to plan for "deadly pandemics and other natural disasters" that can "produce WMD-like effects." It also notes that "environmental destruction, whether caused by human behavior or cataclysmic mega-disasters such as floods, hurricanes, earthquakes, or tsunamis … may overwhelm the capacity of local authorities to respond, and may even overtax national militaries, requiring a larger international response." Like armed attacks, some of the effects of climate change could swiftly kill or endanger large numbers of people and cause such large-scale disruption that local public health, law enforcement, and emergency response units would not be able to contain the threat.

Climate change does not pose an existential risk for a country as large as the United States. Moreover, while Washington, DC, has had its share of storms, the nation's political and military command-and-control center is not as vulnerable to extreme weather events as other parts of the country. However, Hurricane Katrina demonstrated all too well the possibility that an extreme weather event could kill and endanger large numbers of people, cause civil disorder, and damage critical infrastructure in other parts of the country. It would be easy to dismiss that storm's effects as the result of a particularly vulnerable city and an extraordinarily damaging hurricane. But the 2007 IPCC report explicitly warns that coastal populations in North America will be increasingly vulnerable to climate change—and nearly 50 percent of Americans live within fifty miles of the coast. While the Gulf Coast's vulnerability is well known, other densely populated coastal areas are also at risk. For example, a NASA simulation that combined a modest forty-centimeter sea-level rise by 2050 with storm surges from a Category Three hurricane found that, without new adaptive measures, large parts of New

York City would be inundated, including much of southern Brooklyn and Queens and portions of lower Manhattan.[4]

Climate change could, through extreme weather events, have a more direct impact on national security by severely damaging critical military bases, thereby diverting or severely undermining significant national defense resources. This could have a compound effect: in the 2006 Quadrennial Defense Review, the Department of Defense recognized that military assets would likely be called upon in the event of future domestic emergencies. Its consequences might also ripple abroad: for example, Tampa Bay, the site of MacDill Air Force Base and U.S. Central Command (CENTCOM), the center of strategic operations in Iraq, is extremely vulnerable to hurricane damage. A University of South Florida simulation found that the base would likely be inundated if the region were struck by a Category Three hurricane.[5] Other military assets located in Florida are also vulnerable to extreme weather events. U.S. Southern Command (SOUTHCOM), the strategic command for Latin America, is in Miami, another of the cities identified as most vulnerable to hurricane storm damage. In 1992, Hurricane Andrew did such damage to Homestead Air Force Base in Miami that it never reopened. In 2004, damage from Hurricane Ivan kept Pensacola Naval Air Station closed for almost a year. Given the kinds of effects hurricanes have historically had on military bases in the region, it is not farfetched to imagine serious impairment to U.S. national security as Florida sustains further hurricane disasters—and climate change will make such events more severe and potentially more likely.

The effects of climate change on America's neighbors could also be severe, with spillover security effects on the United States. Caribbean countries such as Haiti and Cuba could be hard hit by extreme weather events, contributing to humanitarian disasters as well as the possibility of large-scale refugee flows and state failure. Both Haiti and Cuba have historically used the threat of migration to extract concessions from the United States. In 1980, Fidel Castro forced the United States to accept more than 100,000 Cubans after he encouraged tens of thousands to migrate to Florida during the Mariel

---

[4] Vivien Gornitz and Cynthia Rosenzweig, *Hurricanes, Sea Level Rise, and New York City* (Columbia University, Center for Climate Systems Research, 2006); available at http://www.ccsr.columbia.edu/information/hurricanes/.

[5] Kevin Duffy, "Could Tampa Bay Be the Next New Orleans?" (*Palm Beach Post*, July 9, 2006); available at http://www.palmbeachpost.com/storm/content/state/epaper/2006/07/09/m1a_TAMPA_CANE_0709.html.

boat lift. In 1994, Jean-Bertrand Aristide, in exchange for U.S. intervention to restore him to power, was able to dissuade thousands of Haitians who had constructed makeshift rafts from emigrating to the United States. In the absence of U.S. action to address climate change or support risk reduction, countries in the region could be increasingly tempted to use the threat of migration again.

The United States also faces the likelihood that summer sea ice in the Arctic will be gone by the middle of the century. This will open up the Northern Sea route (north of Russia) and the Northwest Passage (through the Canadian archipelago) to shipping, at least for parts of the year. Both would be attractive for shipping, as they would provide much shorter routes between Europe and Asia than the Panama Canal—4,000 nautical miles less in the case of the Northwest Passage. While this is one of the potential benefits of global warming, the issue threatens to become caught up in interstate disputes over sovereign control over those waters. Canada has claimed the Northwest Passage as internal waters while the United States has asserted they are international waters through which free passage should be permitted. Another concern is contested control of potential petroleum reserves in the area that have heretofore been inaccessible. In the summer of 2007, the Russians raised the stakes by laying claim to the North Pole and the resources underlying it, setting in motion a scramble by other national governments. Though armed confrontation remains unlikely, tensions over territorial waters hearken back to the kinds of border disputes that once led to interstate war.

The United States also has national security interests farther afield, and some of the countries that are vulnerable to climate change may, in particular, be of national security concern to the United States, as sites of U.S. military bases and embassies, allies, potential global or peer competitors, sources of raw materials and/or significant economic partners, sites of major transportation corridors (ports, straits), or places where blowback from events could have an impact on the U.S. homeland. A few specific examples are illustrative.

Indonesia has the world's largest Muslim population—about 88 percent of its 245.5 million people. Some have been radicalized, but most have not. Indonesia is also a fragile democracy and politically unstable with a history of separatist movements. Meanwhile, as an island archipelago with large forest reserves, the country is both vulnerable to climate change and important for climate mitigation. Climate change, through drought conditions or storms, might further destabilize Indonesia, and if the government provided a weak response to a future weather disaster, this could encourage separatists or radicals to challenge the state or launch attacks on Western interests.

Similarly, China is a major economic partner and potential peer competitor. While its authoritarian government currently has a firm grasp on power, rapid social change and widening economic inequality make future instability possible. The country is vulnerable to climate change, as a result of both potential freshwater shortages and storm damage along its densely populated coast. The immense human costs of extreme weather events in cities such as Shanghai and Tianjin could also damage China's industrial production capacity and ports, with knock-on effects on the global economy. An unstable China might also have a less predictable foreign policy.

Other countries with less obvious strategic importance also have large, vulnerable coastal populations. One recent study from the International Institute for Environment and Development found that a tenth of the world's population—634 million people—live in coastal areas that lie between zero and ten meters above sea level.[6] (Storm surges make low-lying coastal areas vulnerable even if sea levels rise only modestly.) Fully 75 percent of those live in Asia. Bangladesh, for example, has 46 percent of its population located in low elevation areas, many of them living in areas less than five meters above sea level. Its capital, Dhaka, with about 12.6 million people, is also one of the most vulnerable cities to flooding. Devastating floods in Bangladesh could send tens of thousands of refugees across the border to India, potentially leading to tension between the refugees and recipient communities in India. In the event of such an emergency, the United States would likely be called upon, given its relief efforts in the region after the 2004 tsunami and the 2005 earthquake in Pakistan. Even if the United States has limited strategic stakes in Bangladesh, support for adaptation measures would still be the right thing to do and much less costly than disaster response.

---

[6] International Institute for Environment and Development, *Climate Change: Study Maps Those at Greatest Risk from Cyclones and Rising Seas* (2007); available at http://www.iied.org/mediaroom/releases/070328coastal.html.

Sub-Saharan Africa is also particularly vulnerable to climate change. While U.S. strategic interests in the region have historically been limited, Africa's growing oil exports to the United States and worries about terrorism have strengthened U.S. interests in the continent.[7] The United States has supported two major antiterrorism efforts in Africa in recent years—one in the Sahel and the other in the Horn of Africa. With the designation of a new U.S. Africa Command (AFRICOM) in 2007, Africa will have an institutional anchor in the military hierarchy. The CNA report concluded that declining food production, extreme weather events, and drought from climate change could further inflame tensions in Africa, weaken governance and economic growth, and contribute to massive migration and possibly state failure, leaving "ungoverned spaces" where terrorists can organize.[8]

The United States has other compelling reasons to be concerned about Africa. The continent's vulnerability to climate change on top of its other problems makes the region especially susceptible to humanitarian disasters with some African governments either unwilling or unable to protect their citizens from floods, famine, drought, and disease. The region is home to a number of unstable regimes and ongoing conflicts, in Somalia, Ethiopia, and the Darfur region of Sudan, with spillover effects on neighboring Chad. Countries in the region are already vulnerable to water shortages, which can exacerbate local grievances and contribute to conflict over scarce resources. Drought conditions (which memorably affected Ethiopia in the 1980s and Somalia in the early 1990s) may be increasingly normal in a world of climate change. Since the United States will be pressured to deploy military forces or at least provide lift and logistic support for large-scale humanitarian emergencies, it has an interest in helping countries minimize the adverse effects of climate change through enhanced local capacity to respond to natural disasters.

These regional examples provide only a partial glimpse of the intersection of U.S. strategic interests, climate vulnerability, and political risk. Nonetheless, they are

---

[7] *More Than Humanitarianism: A Strategic U.S. Approach toward Africa*, Anthony Lake and Christine Todd Whitman, chairs (Council on Foreign Relations Press, 2006); available at http://www.cfr.org/ publication/9302/more_than_humanitarianism.html.

[8] A 2007 RAND report looks at indicators of ungovernability and conduciveness to terrorism in a number of regions. See *Ungoverned Territories: Understanding and Reducing Terrorism Risks*; available at http://www.rand.org/pubs/monographs/2007/RAND_MG561.pdf.

illustrative of the kinds of complex national security challenges the United States will face as climate change intensifies.

PRINCIPLES AND POLICIES FOR CLIMATE AND SECURITY

Responding to the security consequences of climate change will require the United States to support policies that will insulate it as well as countries of strategic concern from the most severe effects of climate change. At the same time, climate policy will also provide the United States with opportunities to improve its relationship with important countries, both rising powers as well as those most vulnerable to environmental damage.

IDENTIFY "NO-REGRETS" POLICIES

The United States should prioritize so-called no-regrets policies, those that it would not regret having pursued even if the consequences of climate change prove less severe than feared.

Domestically, the concentration of human settlements near the coasts justifies many risk-reduction and adaptation policies even if the effects of climate change are modest. Coastal populations are already vulnerable to hurricanes and floods, and policies such as improved building codes make sense irrespective of climate change. Likewise, investments in evacuation and relocation strategies could save lives in the event of terrorist attacks or non-climate-related natural disasters, such as fires or earthquakes. Among other programs that will be beneficial regardless is water conservation, since water scarcity poses a threat to agriculture and human consumption patterns.

Internationally, military-to-military environmental security initiatives (on disaster management, emergency response, and scarce water resources) such as those the U.S. military has sponsored in the Persian Gulf and Central Asia are worthwhile even if the environmental benefits are minimal. U.S. Central Command deputy commander Lieutenant General Michael P. DeLong underscored this point in a 2001 speech: "The United States would not have had access to Central Asia bases to fight the war on terrorism were it not for the relationship established through environmental security programs."

Costs for these conferences were likely in the hundreds of thousands of dollars—a small price to pay. Institutionalizing a series of annual regional conferences at which militaries can discuss natural hazards and disaster preparedness would be among the cheapest investments that the U.S. government could support. For about $100 million, the U.S. government could develop a multiyear program with militaries from Africa, Central Asia, South Asia, Latin America, and the Middle East.[9] At the very least, the meetings could potentially facilitate better ties between militaries (and thereby dampen the possibilities of interstate mistrust). They could also inform the U.S. military about emerging threats, independent of environmental concerns.

No-regrets policies will pay off even if climate change proves less worrisome than many now fear. But given that the worries about climate change are likely to be proved right, policymakers need to go beyond these minimal measures.

DEVELOP POLICIES THAT ADDRESS PROBLEMS IN MULTIPLE DOMAINS

The strongest policies will simultaneously address problems in multiple domains. Policies should address climate security challenges but could also help reduce greenhouse gas emissions, shore up energy security, or provide economic benefits.

Stephen E. Flynn of the Council on Foreign Relations has made a strong case for investments in U.S. physical infrastructure and disaster response capabilities to reduce the potential for catastrophic damage from terrorism, natural disasters, and pandemics. Drawing on a recommendation from the American Society of Civil Engineers, Flynn suggests that an investment in our infrastructure of $295 billion per year for five years will create spillover benefits to the national economy, in the same way the Interstate Highway System did in the 1950s.[10] The United States should support this infrastructure

---

[9] One 2002 study estimated a five-year Central Asia environmental security program (including conferences, staff, exchanges, and small projects) would cost $18 million. Updating for inflation and multiplying by five, a similar initiative today would likely cost $100 million. R.B. Knapp, *Central Asia Environmental Security Technical Workshop: Responding to the Centcom Vision* (Lawrence Livermore National Laboratory, August 1, 2002); available at http://www.llnl.gov/tid/lof/documents/pdf/240886.pdf.

[10] See page xxi in Stephen Flynn, *The Edge of Disaster* (New York: Random House, 2007).

investment program and dedicate a healthy portion to "climate proof" vulnerable infrastructure, particularly in coastal areas.

Another example comes from the Law of the Sea Treaty. As argued earlier, the melting of Arctic ice puts U.S. interests in jeopardy. However, by not ratifying the Law of the Sea Treaty, the United States risks not being party to the adjudicating body that will determine which countries have rights to the region's resources. The Law of the Sea Treaty has been strongly supported by American commercial interests, environmentalists, and the military, all of which see their specific concerns enhanced by ratification. As of this writing, however, a highly motivated few who see treaties as infringements on national sovereignty have stymied final approval. In light of new security concerns from climate change in the Arctic, the U.S. Senate should overcome this inertia and provide its consent to the treaty.

AN OUNCE OF PREVENTION

Reducing risks ahead of time is almost always less costly than responding to disasters after the fact. One estimate from the U.S. Geological Survey and the World Bank suggested an investment of $40 billion would have prevented disaster losses of $280 billion in the 1990s. Between 1960 and 2000, the Chinese spent $3.15 billion on flood control, and averted losses of an estimated $12 billion.[11] Yet the world currently spends too little on adaptive strategies that would reduce climate risk because adaptation has been wrongly perceived as a competitor to mitigation. Supporters of a more robust climate policy have been unenthusiastic about adaptation because they fear it would signal that the world had given up on greenhouse gas emission reductions. This attitude is starting to change, but unless the change is accelerated, the United States and its allies will be forced to expend greater effort later on, including calling upon military assets, to compensate for inadequate risk reduction and disaster response capabilities.

---

[11] See DFID, *Natural Disaster and Disaster Risk Reduction Measures,* December 8, 2005; available at http://www.dfid.gov.uk/pubs/files/disaster-risk-reduction-study.pdf.

The government effort should begin by providing incentives for individuals and firms to reduce risk, particularly through building codes and ensuring that federally funded disaster insurance discourages dangerous coastal settlements. The latter might be done, for example, by limiting government guarantees to rebuild homes and infrastructure that are situated in vulnerable places. The Stern Review, a report by economist Nicholas Stern to the UK government, estimated that the additional resources required to insulate new infrastructure in the United States from climate risk would be on the order of $5 billion to $50 billion per year.[12] Since this estimate includes only new infrastructure, it likely understates the total need. Stephen Flynn's proposal for an infrastructure investment program of $295 billion per year for five years would probably be adequate to "climate proof" critical infrastructure and serve other vital public purposes. While not all of the costs of climate risk reduction will ultimately have to be shouldered by the government, some public resources will be needed, even if these are financed by revenues from a carbon tax or from the auctioning of permits in a cap-and-trade system.

Internationally, there are also scant funds for risk reduction. The World Bank's Global Environmental Facility (GEF) administers two adaptation-related funds for developing countries: the Special Climate Change Fund (SCCF) and the Least Developed Country Fund (LDCF). Together, pledges to GEF adaptation programs cumulatively amount to about $215 million, and although other resources for adaptation exist within the World Bank Group, their scale should be dramatically expanded.[13] Though the United States is a donor to the GEF, the United States has not contributed to either adaptation fund. The Stern Review estimated that it would cost developing countries between $4 billion and $37 billion per year to minimize the climate damage to new investments. Of

---

[12] HM Treasury. *The Stern Review on the Economics of Climate Change*, 2006; available at http://www.hm-treasury.gov.uk/independent_reviews/stern_review_economics_climate_change/sternreview_index.cfm.

[13] In April 2007, for example, the LDCF had total pledges of $115.8 million and the SCCF had pledges of $62 million. Another $50 million was available for the Strategic Priority on Adaptation under the GEF Trust Fund. The December 2007 climate negotiations in Bali will discuss the institutional home for the Adaptation Fund, another funding source derived from a portion of the proceeds from Clean Development Mechanism (CDM) projects. The CDM is one of the Kyoto Protocol's flexibility mechanisms. Global Environmental Facility, *Status Report on the Climate Change Funds* (GEF, June 6-9, 2006); available at http://thegef.org/Documents/Council_Documents/GEF_C28/documents/C.28.4. Rev.1ClimateChange.pdf.

that total, between $2 billion and $7 billion can be expected to come from external finance to cover direct foreign investments vulnerable to climate change. But the balance of developing countries' infrastructure investments needs to be protected, too. Given poor countries' resource constraints, foreign aid should cover at least part of the cost. A modest investment in adaptation in poor countries will likely be much more cost-effective than responding to state failure or humanitarian disasters through military and relief operations.

The United States should take the lead on adaptation by supporting a Climate Change and Natural Disaster Risk Reduction initiative on a scale similar to President George W. Bush's Emergency Plan for AIDS Relief. The president's AIDS plan delivered $15 billion over five years through a combination of bilateral programs and support for the Global Fund to Fight AIDS, TB, and Malaria. Based on the initial estimates of the adaptation costs for poor countries, climate risk reduction should have at least that level of support, divided between bilateral and multilateral programs. The bulk of this support should finance adaptation programs by vulnerable governments. It should go beyond the protection of new infrastructure and include agricultural research and planning for emergencies. The previously mentioned military-to-military workshops for disaster management should form part of this effort, too. The United States should take advantage of the creation of AFRICOM to create a multiagency African Risk Reduction Pool with a budget of at least $100 million per year. AFRICOM may develop new ways of incorporating climate and other environmental concerns into conflict prevention. It may also serve as a model for interagency coordination that is applicable to other regions.

AFRICOM is already structured to have a State Department official as the deputy. There is a danger, however, that the mission will be conceived narrowly as capturing and killing terrorists. While important, another component should be conflict prevention to address the underlying causes of political instability, including the potential for climate change to contribute to refugee crises and water and resource scarcity, among other problems. For conflict prevention, most environmental adaptation programming will be development work rather than traditional security initiatives, necessitating greater on-the-ground coordination between civilian and military agencies.

The idea of a risk-reduction pool is based on the African Conflict Prevention Pool (ACPP), a collaborative effort by the United Kingdom's Department for International Development (DFID), the Ministry of Defence (MOD), and the Foreign and Commonwealth Office (FCO), which is equivalent to the U.S. State Department. The three agencies pool their funds for conflict prevention. The ACPP, with a subsidy from the UK Treasury, has a budget of more than £60 million per year ($120 million). Funds are administered by all three agencies in a "joined-up" field operation where interagency teams collaborate on a common strategy. Depending upon their area of expertise and comparative advantage, individual agencies draw down resources for different purposes (such as security sector reform, demobilization of soldiers, efforts to control small arms, and programs addressing the economic and social causes of conflict). This model offers great potential for enhanced interagency collaboration in the field and can minimize duplicative programming. The U.S. version, likewise, should draw upon a broader number of civilian agencies, including the Environmental Protection Agency (EPA), the Department of Agriculture, NASA, the U.S. Army Corps of Engineers, and NOAA (National Oceanic and Atmospheric Administration). The pool should finance a number of climate security initiatives, including early warning systems of extreme weather events as well as investments in coastal defenses, water conservation, dispute settlement systems, and drought-resistant crops. Whatever climate changes come to pass, these measures would be designed to minimize the potential consequences for political stability and social strife.

Even a successful portfolio of risk-reduction and conflict prevention strategies will experience an occasional failure. When crises do strike, the pool approach would facilitate rapid response and better integration of military and civilian efforts to move quickly from emergency to postcrisis. The pool should finance contingency plans for humanitarian relief operations and purchase some relief supplies in the event of crises, including surplus grains from African farmers. While African countries may resist a heavy U.S. footprint, the United States should consider some pre-positioning of lift capabilities and ground transport for emergency situations either at the main base in Africa, once established, or distributed throughout the region as needed. The Africa example is a model of what potentially could be extended to other vulnerable regions.

At the domestic level, the absence of a federal policy on climate change has paralyzed more proactive efforts to insulate the United States from climate risks, and this extends even to efforts to document the problem. As of October 2007, the House of Representatives had appropriated $50 million for an innovative two-year Commission on Climate Change Adaptation and Mitigation for fiscal year 2008. The Senate had not appropriated any support for the commission. An important complement to the ongoing National Intelligence Estimate on climate and security, the commission would allow the EPA to fund studies by a number of federal agencies. Studies could also consider the cost-effectiveness of different policy remedies, including improved sea defenses, building codes, emergency response plans, and even relocation strategies. Congress should fully fund the commission's activities. Research should focus on whether and where extreme weather events could cause localized humanitarian crises, divert or disable national security instruments, or contribute to regional disorder.

Internationally, the first priority should be to generate more precise estimates of adaptation costs. The United States, perhaps acting in coordination with the IPCC or World Bank, should take existing studies of coastal areas vulnerable to climate change and evaluate which strategies are likely to yield the most damage reduction at the least cost. A similar analysis should be conducted for food production and freshwater availability. A global assessment from the Bank might identify countries most vulnerable to climate change without regard to their underlying geopolitical importance. A U.S. risk assessment might be more targeted, focusing on countries that are of more obvious national security concern to the United States. The National Intelligence Council is preparing an analysis on climate change and national security that may provide a first assessment of this challenge. More global studies would have the advantage of pooling expertise and potentially identifying areas of non-obvious security significance.

Analysis and projections should be supplemented with more sophisticated real-time information on changing climate conditions. While meteorological information about the United States is extensive, satellite coverage in other parts of the globe is patchy. One asset that would be valuable, particularly in the African context, is the High

Altitude Airship (HAA), an unmanned blimp that can be positioned for months at a time to monitor weather systems and provide more continuous surveillance than a satellite. Unfortunately, funding for an HAA prototype from the Missile Defense Agency has been cut in recent years and is scheduled for elimination in 2008. For this worthwhile program to continue, Congress should appropriate $100 million over the next three fiscal years to ensure that the prototype is ready by the end of fiscal year 2010.

MITIGATION POLICY AS DIPLOMACY

While risk reduction is essential, climate damages are likely to exceed most governments' adaptive capacities unless a major reduction in greenhouse gases takes place before the mid-twenty-first century. For example, the IPCC reports that by 2050, three coastal deltas—the Nile, the Mekong, and Ganges-Brahmaputra—will be extremely vulnerable to climate change, meaning that more than a million people could be displaced.[14] And just as many adaptation policies have clear national security dimensions, so do many possible mitigation initiatives. Consider three cases that illustrate this: China, India, and Indonesia.

Engagement remains the most important strategy to encourage China to become a status quo power and reduce the risk that China's rise leads to confrontation between the great powers. Climate policy provides a valuable avenue for such engagement. While advanced industrialized countries bear historic responsibility for existing concentrations of greenhouse gases, China will be increasingly fingered as a climate culprit in the future. This will create a common interest between the United States and China in avoiding world condemnation for being "climate villains." Enlightened climate diplomacy could build on that common interest to improve U.S.-China relations.

At the same time, climate change could also possibly become a wedge issue in the U.S.-China relationship. For example, a climate bill currently before Congress would

---

[14] R.J. Nicholls, et al. "Coastal systems and low-lying areas." *Climate Change 2007: Impacts, Adaptation and Vulnerability. Contribution of Working Group II to the Fourth Assessment Report of the Intergovernmental Panel on Climate Change* (Cambridge University Press, Cambridge, UK, 2007); available at http://www.gtp89.dial.pipex.com/06.pdf.

allow the president, if he or she deems a country's climate efforts to be inadequate, to impose tariff-like fees on carbon-intensive imports such as steel beginning in 2019. Such legislation, if passed, would probably be used against China, adding to existing frictions over trade, intellectual property, and the level of China's currency. So just as climate change presents an opportunity to solidify relations with China, so too does it present the possibility of new tensions in the relationship. Deft handling of the climate dimension of the U.S.-China relationship could have profound implications.

Once the United States joins other rich countries in adopting a domestic regime to control carbon emissions, climate change will become an important part of the global rules–based order. Whether China chooses to remain engaged depends on whether it can meet its perceived needs inside the system. A climate policy that induces China to join the rules-based global regime for dealing with global warming—independent of the fine details of that policy—would contribute to the broader project of cementing China's commitment to the world order, which in turn could create payoffs in building a positive security relationship. At the same time, clumsy handling of climate issues could sour relations more broadly, damaging American security interests well beyond the climate sphere.

The same is true of India. While the 2006 nuclear agreement with India was designed to reduce the threat of nuclear proliferation,[15] from a security-oriented climate perspective, the nuclear deal also has the potential to restrain the country's greenhouse gas emissions. David G. Victor of Stanford University and the Council on Foreign Relations estimates that if India were to build twenty gigawatts of nuclear power as envisioned in the 2006 agreement, this could save 145 million tonnes per year of carbon dioxide emissions that would otherwise have been belched from coal-generating plants.[16]

As part of this broader strategy of geopolitically informed climate policy, the United States should make sure that enhancing formal participation by China and India in

---

[15] Michael A. Levi and Charles D. Ferguson, *U.S.-India Nuclear Cooperation* (Council on Foreign Relations, June 2006;); available at http://www.cfr.org/content/publications/attachments/USIndiaNuclear CSR.pdf.

[16] David G. Victor, "The India Nuclear Deal: Implications for Global Climate Change," testimony before the U.S. Senate Committee on Energy and Natural Resources (Stanford University Program on Energy and Sustainable Development, July 18, 2006); available at http://pesd.stanford.edu/ publications/india_nuclear_deal.

important global institutions is a part of its climate change mitigation strategy. In particular, it should promote closer engagement between China and India and the International Energy Agency (IEA), a body that currently excludes both countries from its membership. The IEA is an important organization for building trust and cooperation among energy consumers. It will also be increasingly significant in helping countries reduce greenhouse gases.[17] Already, the IEA has memoranda of understanding with both countries to enhance cooperation on climate change; were the U.S. government to support the deepening of these ties with an eye toward eventual membership, it would help advance climate goals while further integrating China and India into the rules-based global order.

Indonesia provides another example. Indonesia is a major player in climate change because of deforestation, which releases carbon stored in plant matter and the soil: deforestation and forest fires in Indonesia helped make it the third-largest contributor of greenhouse gases behind the United States and China. Paying Indonesia to keep its forests would likely be a much cheaper way for rich countries to avoid emitting greenhouse gases than retrofitting existing industrial infrastructure or seeking a rapid change in transportation fuels. But there is a security angle here, too. Indonesia's political instability has fostered terrorist groups that may have global ambitions. Managing forestry payments deftly could help to solidify Indonesia's social order and discourage radicals. In Aceh, for example, the provincial government, led by a former rebel, is seeking support for avoided deforestation as a means of persuading former separatists to protect the forests and refrain from picking up their guns; providing him with the resources he seeks could mitigate both climate change and separatism. Similar win-win opportunities may exist in other strategically important countries, including Brazil and the Democratic Republic of the Congo.

The idea of compensating other countries for avoided deforestation has gained attention in recent years, spurred by a proposal from Costa Rica and Papua New Guinea on behalf of forest-rich countries at the 2005 Conference of Parties (COP) in Montreal. However, the Kyoto Protocol, largely because of worries about problems monitoring and

---

[17] IEA membership is linked to membership in the Organization for Economic Cooperation and Development (OECD). The OECD has ongoing discussions about making both China and India members, though the lack of democracy in China may prove an impediment to formal membership.

charting actual savings, did not allow avoided deforestation to generate tradable emissions credits. Thus, countries can get paid for replanting forests but not for preventing them from being cut down in the first place. At the 2007 G8 Summit, the World Bank achieved agreement on a $250 million pilot project for avoided deforestation in five countries.[18] The Bank is now seeking funding for the pilot; the program's official launch is supposed to take place at the climate negotiations in Bali in December 2007. However, if countries like Indonesia are to benefit and if savings from avoided emissions from forestry are to materialize, the United States must play an active role in addressing the remaining technical issues and ensure the pilot program is fully funded.[19] At the same time, the U.S. has an opportunity to shape future climate negotiations by insisting that credits from avoided deforestation be included in a successor agreement to the Kyoto Protocol.

---

[18] Avoided deforestation is now also referred to as reduced emissions from deforestation and degradation (REDD). The World Bank estimates that the pilot program could result in about 40 million tonnes in avoided carbon dioxide emissions between 2008 and 2012, conserving about 100,000 hectares of forest. The World Bank has proposed an ambitious Global Forest Alliance (GFA), partnering with large environmental nongovernmental organizations like the Nature Conservancy to implement the program, the so-called Forest Carbon Partnership Facility.

[19] Among the main technical issues that need to be determined are how to develop baselines for emissions reductions and how to compensate states and individual forest owners for their actions.

INSTITUTIONAL REFORMS

The importance of climate policy to national security demands that it receive much greater prioritization across the U.S. federal government. In the current administration, climate policy is largely run by two players: the head of the White House Council on Environmental Quality cooperates with the senior climate negotiator at the State Department in the Bureau of Oceans and International Environmental and Scientific Affairs. Other players in the federal government have largely been sidelined. There are few efforts to integrate climate concerns into top-level decision-making.

Several positions created during the 1990s but abolished in recent years could be useful templates for the future. A special assistant position on climate change, focused only on climate rather than on the broad range of environmental questions that fall under the purview of the director of the Council on Environmental Quality, was tasked to oversee interagency cooperation. The NSC also had a senior director for environmental affairs, a position that was later eliminated and folded under International Trade, Energy, and the Environment. The Department of Defense created a deputy undersecretary of defense for environmental security tasked to deal both with the environmental footprint of the military and the emerging security concerns associated with environmental harms and natural hazards. The military's environmental impact was later subsumed into the portfolio of the deputy undersecretary for installations and environment; the substantive policy focus was dropped.

Given the strong links between climate change and security, the rebuilding of a cadre of officials focused on climate should begin at the Pentagon. A new deputy undersecretary of defense position for environmental security (under the broader mandate of OSD's policy office) should be created to redress the insufficient institutionalization of climate and environmental concerns in DOD decision-making.[20] With a small staff of roughly two dozen people, that office could provide constant attention to the strategic dimensions of emerging environmental security threats and champion specific proposals

---

[20] Environmental cleanup and conservation should remain in a separate office on the logistics and installations side of DOD.

like the military-to-military workshops, the African Risk Reduction Pool, and investment in the High-Altitude Airship. At the same time, the environmental security outfit could ensure that other offices charged to deal with homeland security look beyond terrorism to consider environmental threats like extreme weather events. Concerns about emerging environmental harms should also be integrated into the planning and operations of the regional combatant commands. Unless these issues are perceived to be a priority by DOD leadership at the highest levels, regional commanders might treat environmental security as solely the preserve of this small new office. That would be a mistake. The next president can ensure the issue gets the priority it deserves by integrating climate security concerns centrally into its National Security Strategy.[21]

It would be counterproductive, though, to treat climate security concerns solely or even primarily with the traditional tools of national defense. These instruments may not be best suited for the purposes of reducing the vulnerability of countries to natural hazards made worse by climate change. While greater integration of the military into coordinated disaster planning will be useful and necessary, there is a tendency to confuse national defense with what the military can do. Adaptation policies, both at home and abroad, will likely be supported by non-Defense Department agencies.

Mobilization of all the tools in the U.S. government's arsenal will require high-level attention to climate change among White House officials who lead the interagency process. Since climate change is an issue that straddles domestic and international domains, neither the National Security Council nor the Domestic Policy Council is equipped on its own to develop a coherent response across the federal government. The president should direct the leadership of both agencies to work together to clarify responsibilities and coordination mechanisms so that climate security concerns do not fall through the cracks.

Leadership from the White House could take several forms. The president could re-create a senior director position at the NSC and a small number of supporting directors to deal with climate change and the environment. Given the cross-cutting nature of the

---

[21] That, in turn, will set the stage for broader climate security concerns to cascade down, as they should, into other planning efforts like the Quadrennial Defense Review and Theater Security Cooperation Plans. Theater Security Cooperation Plans are documents that combatant commands use to coordinate regional security cooperation during peacetime.

issue, that senior director should be appointed to the NSC, the Council on Environmental Quality, and the National Economic Council. But since these officials would not control agency budgets, additional senior NSC staff might be needed to ensure that the security dimensions of climate policy get sufficient attention. One possibility is to create a deputy national security adviser (DNSA) position for sustainable development, tasked to oversee foreign assistance, humanitarian issues, pandemic disease, and also emerging environmental threats like climate change. The DNSA and the senior director for environmental affairs at the NSC would be well placed to guide the interagency process.

Even with these recommendations, the links between climate and security still might not get sufficient attention. One additional institutional change could overcome this problem by placing climate change closer to the president. Special advisers to the president with some budgetary authority can be especially effective. President Bush created positions for a global AIDS coordinator and a director of foreign assistance. Presidents have long had drug czars. President Bill Clinton had a special assistant for climate change. Re-creation of this post with actual budgetary authority would go a long way to driving interagency coordination and ensuring more coherence between the national security pieces of this problem and those related to energy and environmental protection.

Congress should adopt a more limited agenda of institutional change to increase the visibility of climate security. Climate security touches on the potential jurisdiction of a number of different committees, and given the fractious nature of the legislative branch, it may be difficult to channel climate security concerns in a way that boosts their salience.

The House Select Committee on Energy Independence and Climate Change was created to focus climate and energy policy, but its mandate is set to expire at the end of October 2008. It may be helpful to make this committee permanent while tasking it to identify appropriate policies to reduce U.S. vulnerability to climate security risks. Since this committee is new and has no legislative authority, it is unclear if it will be a useful vehicle even before it expires. Congress should wait to evaluate its success before making it permanent.

Congressional oversight, though, will be important regardless. The assessment of climate and security that is being prepared by the National Intelligence Council (NIC) should be provided to the relevant committees in Congress, including the House and Senate Committees on Appropriations, Armed Services, and the Select Committees on Intelligence. Congress should ask the NIC to provide regular updates on climate and security risks. At the very least, it would be useful for Congress to get updated assessments after the release of important peer-review reports of climate science, like the IPCC assessments or those the president may request from the National Academy of Sciences.

CONCLUSION

Until recently, the debate about climate change has emphasized how large the economic consequences are, how these compare to the costs of action, and whether the United States or other nations can afford to address the issue. Extreme weather events such as Hurricane Katrina, the fires in Greece, and the floods in Africa and Asia suggest a different way of thinking about the issue. The macroeconomic costs of Hurricane Katrina were minimal in the context of a large and resilient U.S. economy, but the human and political consequences were significant and painful. Whether or not Katrina was linked to global warming, climate change will likely yield more of these kinds of episodes, which are characterized by concentrated costs to particular places and people, leading to severe local impacts and cascading consequences for others.

The concentrated impacts of climate change will have important national security implications, both in terms of the direct threat from extreme weather events as well as broader challenges to U.S. interests in strategically important countries. Domestically, extreme weather events made more likely by climate change could endanger large numbers of people, damage critical infrastructure (including military installations), and require mobilization and diversion of military assets. Internationally, a number of countries of strategic concern are likely to be vulnerable to climate change, which could lead to refugee and humanitarian crises and, by immiserating tens of thousands, contribute to domestic and regional instability.

Climate policy should seek to avoid the worst consequences of global warming. It should start with no-regrets measures that make sense even if the consequences of climate change prove less than severe. These include coastal protection at home and support for military-to-military environmental security conferences overseas. In addition, the United States should support policies that simultaneously address multiple problems, such as those that reduce security risks but also provide economic benefits—investments in infrastructure, for example. The United States must also recognize that the existing concentration of greenhouse gases guarantees that some climate change is inevitable. U.S. policies should thus support risk reduction and adaptation at home and abroad.

Specific adaptation policies that could be supported are early warning systems, building codes, emergency response plans, coastal defenses, and evacuation and relocation schemes.

While risk-reduction programs are a necessary component of a climate policy that addresses national security, the United States and the world will need to move to a decarbonized energy future before century's end—it is widely agreed that a major push to support new technologies to lower greenhouse gas emissions and sequester carbon is essential. But policymakers must recognize that mitigation policies involve not only costs but also opportunities to strengthen national security. A new compact on clean energy technology transfer to China and India would bolster support for the rules-based global order that the United States has nurtured since World War II. An avoided deforestation scheme, particularly in strategically important countries such as Indonesia, could not only reduce greenhouse gas emissions but also support stability and conflict resolution. Finally, for these policy recommendations to have traction, institutional reform is needed. To give voice to climate and security concerns, several new positions should be created across the executive branch—in the Department of Defense, in the National Security Council, and in the Office of the President.

The policy proposals presented here are illustrative rather than exhaustive, but they have the potential to strengthen national security by reducing U.S. vulnerabilities to climate change at home and abroad, securing and stabilizing important partners, and contributing to other goals such as energy security and industrial revitalization. In a world of new security challenges, forging a climate policy along these lines must be a national priority.

ABOUT THE AUTHOR

Joshua W. Busby is an assistant professor at the Lyndon B. Johnson School of Public Affairs and is affiliated with the Robert S. Strauss Center for International Security and Law, both at the University of Texas at Austin. In 2004, Dr. Busby and Nigel Purvis of the Brookings Institution contributed a paper for the UN High-Level Panel on Threats, Challenges, and Change titled "The Security Implications of Climate Change for the UN System." A forthcoming article, "Who Cares About the Weather? Climate Change and U.S National Security," will appear in *Security Studies*.

Dr. Busby has been a research fellow at the Brookings Institution, Harvard University's Belfer Center for Science and International Affairs, and Princeton University's Woodrow Wilson School of Public and International Affairs. His work has appeared in *International Studies Quarterly* and *Current History*, among other publications. He served in the Peace Corps in Ecuador from 1997 to 1999. Dr. Busby is a term member of the Council on Foreign Relations and a member of the International Institute for Strategic Studies. He has a BA from both the University of North Carolina–Chapel Hill and the University of East Anglia, where he was a British Marshall Scholar, and he received his MA and PhD from Georgetown University.

ADVISORY COMMITTEE FOR

CLIMATE CHANGE AND NATIONAL SECURITY

Kent Hughes Butts
U.S. ARMY WAR COLLEGE

Kurt M. Campbell, Chair
CENTER FOR A NEW AMERICAN SECURITY

Helima L. Croft
LEHMAN BROTHERS

John Gannon
FORMER CHAIRMAN, NATIONAL
INTELLIGENCE COUNCIL

Lukas Haynes
MERTZ GILMORE FOUNDATION

Paul F. Herman Jr.
NATIONAL INTELLIGENCE COUNCIL

Jeff Kojac
LIEUTENANT COLONEL
U.S. MARINE CORPS

Marc A. Levy
CENTER FOR INTERNATIONAL EARTH
SCIENCE INFORMATION NETWORK,
COLUMBIA UNIVERSITY

Meg McDonald
ALCOA, INC.

Alisa Newman Hood
WHITE & CASE LLP

Stewart M. Patrick
CENTER FOR GLOBAL DEVELOPMENT

Joseph Wilson Prueher
ADMIRAL, USN (RET.)

Nigel Purvis
UNITED NATIONS FOUNDATION

P.J. Simmons
SEA STUDIOS FOUNDATION

R. James Woolsey
FORMER DIRECTOR OF CENTRAL
INTELLIGENCE

MISSION STATEMENT OF THE MAURICE R. GREENBERG
CENTER FOR GEOECONOMIC STUDIES

Founded in 2000, the Maurice R. Greenberg Center for Geoeconomic Studies at the Council on Foreign Relations works to promote a better understanding among policymakers, academic specialists, and the interested public of how economic and political forces interact to influence world affairs. Globalization is fast erasing the boundaries that have traditionally separated economics from foreign policy and national security issues. The growing integration of national economies is increasingly constraining the policy options that government leaders can consider, while government decisions are shaping the pace and course of global economic interactions. It is essential that policymakers and the public have access to rigorous analysis from an independent, nonpartisan source so that they can better comprehend our interconnected world and the foreign policy choices facing the United States and other governments.

The center pursues its aims through:

- Research carried out by Council fellows and adjunct fellows of outstanding merit and expertise in economics and foreign policy, disseminated through books, articles, and other mass media;
- Meetings in New York, Washington, DC, and other select American cities where the world's most important economic policymakers and scholars address critical issues in a discussion or debate format, all involving direct interaction with Council members;
- Sponsorship of roundtables and Independent Task Forces whose aims are to inform and help to set the public foreign policy agenda in areas in which an economic component is integral; and
- Training of the next generation of policymakers, who will require fluency in the workings of markets as well as the mechanics of international relations.

COUNCIL SPECIAL REPORTS
SPONSORED BY THE COUNCIL ON FOREIGN RELATIONS

Planning for a Post-Mugabe Zimbabwe
Michelle D. Gavin; CSR No. 31, October 2007
A Center for Preventive Action Report

The Case for Wage Insurance
Robert J. LaLonde; CSR No. 30, September 2007
A Maurice R. Greenberg Center for Geoeconomic Studies Report

Reform of the International Monetary Fund
Peter B. Kenen; CSR No. 29, May 2007
A Maurice R. Greenberg Center for Geoeconomic Studies Report

Nuclear Energy: Balancing Benefits and Risks
Charles D. Ferguson; CSR No. 28, April 2007

Nigeria: Elections and Continuing Challenges
Robert I. Rotberg; CSR No. 27, April 2007
A Center for Preventive Action Report

The Economic Logic of Illegal Immigration
Gordon H. Hanson; CSR No. 26, April 2007
A Maurice R. Greenberg Center for Geoeconomic Studies Report

The United States and the WTO Dispute Settlement System
Robert Z. Lawrence; CSR No. 25, March 2007
A Maurice R. Greenberg Center for Geoeconomic Studies Report

Bolivia on the Brink
Eduardo A. Gamarra; CSR No. 24, February 2007
A Center for Preventive Action Report

After the Surge: The Case for U.S. Military Disengagement from Iraq
Steven N. Simon; CSR No. 23, February 2007

Darfur and Beyond: What Is Needed to Prevent Mass Atrocities
Lee Feinstein; CSR No. 22, January 2007

Avoiding Conflict in the Horn of Africa: U.S. Policy Toward Ethiopia and Eritrea
Terrence Lyons; CSR No. 21, December 2006
A Center for Preventive Action Report

Living with Hugo: U.S. Policy Toward Hugo Chávez's Venezuela
Richard Lapper; CSR No. 20, November 2006
A Center for Preventive Action Report

Reforming U.S. Patent Policy: Getting the Incentives Right
Keith E. Maskus; CSR No. 19, November 2006
A Maurice R. Greenberg Center for Geoeconomic Studies Report

Foreign Investment and National Security: Getting the Balance Right
Alan P. Larson, David M. Marchick; CSR No. 18, July 2006
A Maurice R. Greenberg Center for Geoeconomic Studies Report

Challenges for a Postelection Mexico: Issues for U.S. Policy
Pamela K. Starr; CSR No. 17, June 2006 (web-only release) and November 2006

U.S.-India Nuclear Cooperation: A Strategy for Moving Forward
Michael A. Levi and Charles D. Ferguson; CSR No. 16, June 2006

Generating Momentum for a New Era in U.S.-Turkey Relations
Steven A. Cook and Elizabeth Sherwood-Randall; CSR No. 15, June 2006

Peace in Papua: Widening a Window of Opportunity
Blair A. King; CSR No. 14, March 2006
A Center for Preventive Action Report

Neglected Defense: Mobilizing the Private Sector to Support Homeland Security
Stephen E. Flynn and Daniel B. Prieto; CSR No. 13, March 2006

Afghanistan's Uncertain Transition From Turmoil to Normalcy
Barnett R. Rubin; CSR No. 12, March 2006
A Center for Preventive Action Report

Preventing Catastrophic Nuclear Terrorism
Charles D. Ferguson; CSR No. 11, March 2006

Getting Serious About the Twin Deficits
Menzie D. Chinn; CSR No. 10, September 2005
A Maurice R. Greenberg Center for Geoeconomic Studies Report

Both Sides of the Aisle: A Call for Bipartisan Foreign Policy
Nancy E. Roman; CSR No. 9, September 2005

Forgotten Intervention? What the United States Needs to Do in the Western Balkans
Amelia Branczik and William L. Nash; CSR No. 8, June 2005
A Center for Preventive Action Report

A New Beginning: Strategies for a More Fruitful Dialogue with the Muslim World
Craig Charney and Nicole Yakatan; CSR No. 7, May 2005

Power-Sharing in Iraq
David L. Phillips; CSR No. 6, April 2005
A Center for Preventive Action Report

Giving Meaning to "Never Again": Seeking an Effective Response to the Crisis in Darfur and Beyond
Cheryl O. Igiri and Princeton N. Lyman; CSR No. 5, September 2004

Freedom, Prosperity, and Security: The G8 Partnership with Africa: Sea Island 2004 and Beyond
J. Brian Atwood, Robert S. Browne, and Princeton N. Lyman; CSR No. 4, May 2004

Addressing the HIV/AIDS Pandemic: A U.S. Global AIDS Strategy for the Long Term
Daniel M. Fox and Princeton N. Lyman; CSR No. 3, May 2004
Cosponsored with the Milbank Memorial Fund

Challenges for a Post-Election Philippines
Catharin E. Dalpino; CSR No. 2, May 2004
A Center for Preventive Action Report

Stability, Security, and Sovereignty in the Republic of Georgia
David L. Phillips; CSR No. 1, January 2004
A Center for Preventive Action Report

To purchase a printed copy, call the Brookings Institution Press: 800-537-5487.
Note: Council Special Reports are available to download from the Council's website, CFR.org.
For more information, contact publications@cfr.org.

www.ingramcontent.com/pod-product-compliance
Lightning Source LLC
Chambersburg PA
CBHW051346290326
41933CB00042B/3309